EYES AND EARS

BY MARK J. RAUZON

For Eric Knudtson

Acknowledgments

Special thanks to the following people for the use of their photographs: Jacquelyn Gerhart and Dr. Mindy Weinstein for the microphotograph of spider eyes on page 15; and Dr. Bart Smith for the photograph of Mark Rauzon's eye on the back jacket.

LOTHROP, LEE & SHEPARD BOOKS   NEW YORK

ASIAN ELEPHANT ▲

JACKSON'S CHAMELEON ▼

MACAWS

HOUSE CAT

All animals need to know what is going on around them to survive. They get this information through their five senses: sight, hearing, touch, taste, and smell. All five senses are important in order to live, but most animals rely more on their eyes and ears to find food and mates and to avoid danger.

▲ FOREST IGUANA   BALD EAGLE ▶

EYES

Eyes sense light, motion, and color. Almost all animals have eyes, but each sees in its own way. Animals such as lizards and snakes can't see far, but their eyes can sense movement very well. Others, including this eagle, can spot a snake in the grass from a mile away. Eagles and hawks need to see long distances, since they hunt from high overhead. They hunt during the day because their eyes need a great deal of light to work properly. They can't see nearly as far or as clearly in the dark as they do in daylight.

◀ **OCELOT KITTEN**

▲ **RACCOON**

No animal can see in total darkness, but many can see in very dim light. Raccoons, ocelots, and other nighttime hunters see well in the dark. Their eyes are large to let in as much light as possible. At night, an animal's eyes reflect direct light and shine like glowing embers.

Most animals are color-blind. They see the world in black and white and shades of gray. Only a few kinds of animals, including insects, birds, and humans, see colors. Color can be sensed only if there is plenty of light, so animals don't see color at night even if they do in the day.

Colors help animals attract mates, find food, and avoid enemies. They help animals know when fruits are ripe and which flowers are sweet. Brightly colored amphibians and insects are often poisonous or taste bad. Their colors warn predators not to eat them.

PSEUDO SPHINX MOTH CATERPILLAR

SPOTTED OWL AND FIELD MOUSE ▲   **LLAMA ▶**

The eyes of most hunting animals, such as this spotted owl, are set close together and look straight ahead so they can see to grab the prey they are chasing. Monkeys' and humans' eyes also face forward so that they can see clearly to use their hands.

But the eyes of most plant-eating animals are set far apart on either side of their heads so that they can watch for predators while grazing. This llama can see almost all the way around itself without having to turn its head.

Hippos spend most of their lives in the water. Their eyes are set in bumps on top of their heads so that they can see above water while they are swimming. Alligators, crocodiles, and frogs have eyes atop their heads, too.

Eyes are very delicate. They need to be protected from sharp objects and always kept moist. Almost all kinds of animals have

HIPPOPOTAMUS

CLOSE-UP OF SHARK EYE

BLACK-TIPPED REEF SHARK

eyelids that they can close over their eyes to shield them from injury and drying out. Fish, such as this shark, are in no danger of having dry eyes, so they don't have eyelids. Fish must sleep with their eyes wide open.

Flatfish, including flounders, spend most of their lives grubbing for food on the ocean floor. Their eyes are set close together on the top of their heads.

BANANA SLUG

Most kinds of animals have two eyes, located in their heads, but a few need more than a single pair. This banana slug has four eyes, mounted on stalks, so that it can peer over and under leaves at the same time. Spiders have *eight* eyes. They can search for food and keep watch for danger in all directions at once.

▲ MICROPHOTO
OF SPIDER EYES

◀ **INDIAN RHINOCEROS**   **BELUGA WHALE** ▲

EARS

Ears, like eyes, help animals sense what is going on around them. Ears hear vibrations called sounds. Just as animals don't all see the same way, they don't hear the same way.

Some animals, including humans, rely more on their eyes than on their ears. But most animals hear better than they see. Indian rhinos can't see far through the tall grasses they live in, but their ears let them know when hunting tigers are approaching. Eyes aren't much help, either, to a lonely beluga whale in the black waters of the Arctic Ocean, but its ears let it hear other whales singing miles away.

◄ BLACK-TAILED JACKRABBIT　　　　　　　　**▲ RED FOX**

A jackrabbit's huge, cupped outer ears can swivel around so it can listen in any direction without turning its head. If the rabbit is very alert and hears an enemy coming, it may have time to run away before it is seen.

Unfortunately for jackrabbits, foxes and wolves also have cupped outer ears, to help them pinpoint the location of their prey.

Birds have no outer ears, just small ear holes hidden by feathers. Because it has no outer ears to catch sounds, this burrowing owl must turn its head to listen in different directions.

BURROWING OWL

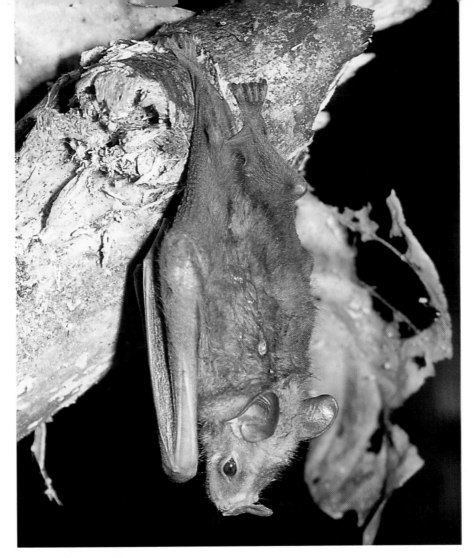

LEAF-NOSED BAT

Some bats use their funnel-shaped ears to help them hunt through the night sky. This leaf-nosed bat makes high-pitched sounds that bounce off objects and uses the echo of its own chirp to locate insects to eat.

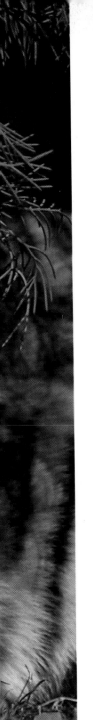

◄ **SUMATRAN TIGER**　　　　　▲ **YELLOW-RUMPED CACIQUE**

Many kinds of animals make sounds to communicate with one another.

Coyotes howl to call the pack together for a hunt. Birds sing songs to attract mates and establish territories. Tigers purr when they are content, growl when they are annoyed, and their mighty roars broadcast their presence. A male elk "roars," too. Its loud bugling warns other males to leave its territory or risk a fight.

CRESTED CARACARA

ISLAND FOX

GIRAFFE

NEW ZEALAND FUR SEAL PUP

All over the world, animals use their own unique kinds of eyes and ears to see and hear what is going on around them and to help them survive.